PRODUCTION SYSTEMS AND LINE BALANCING IN CLOTHING INDUSTRY

Introduction

In general, most of the production systems employed in clothing factories are derived from the following manual or mechanical systems. Each production system has its own specific operational characteristics in terms of:

- Supervision
- Labour
- Quality Control
- Productivity
- Throughput time
- Layout

These factors are examined in this chapter

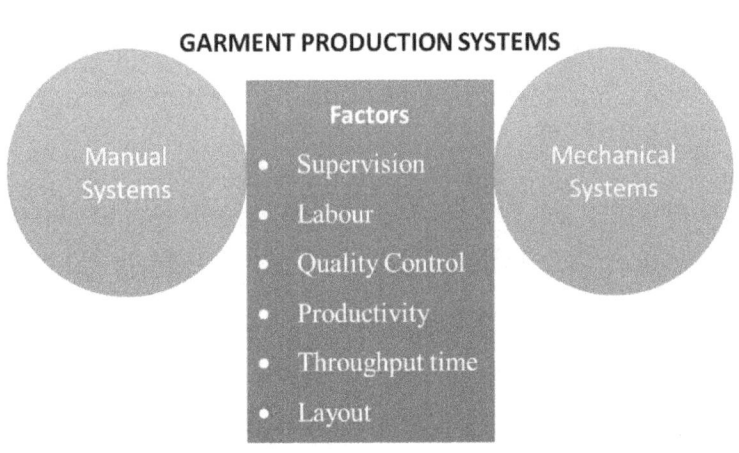

MANUAL SYSTEMS

Individual System/Make Through/Whole Garment system

This is essentially the traditional method of production whereby the entire garment is assembled by one operator. In men's bespoke wear, it is not uncommon for a tailor to perform nearly every operation required to make the garment, including machining, hand work and pressing.

With this production system the operator would be given a bundle of cut work and would proceed to sew it according to his or her own method of work. Of necessity, the labour required by this system must be highly skilled and versatile, a combination which is becoming exceedingly rare and increasing expensive. The characteristics of the making through system are:

- **Supervision**: The requirements are minimal because the operators decide on their own working methods. At the most a supervisor would provide bundles of work together with some explanation, and then remove the finished work when ready.

- **Labour:** As highly skilled labour is used for the simplest of operations, costs are relatively high when compared with other systems.

- **Quality Control:** Obviously this is necessary, but not at the same level as required for garments, produced by semi-skilled or unskilled labour.

- **Productivity:** Due to the lack of specialization, this system is not conducive to high levels of productivity.
- **Throughput time:** This is reduced to a minimum because there is no real necessity for reserves of work except for that actually being sewn by the operator.
- **Layout:** A convenient arrangement of machinery etc, is all that is required.

This type of system is effective when a very large variety of garments have to be produced in extremely small quantities. A typical application would be in the sewing room of a boutique, which produces its own merchandise.

WHOLE GARMENT PRODUCTION SYSTEM

There are two types of Whole Garment Production Systems

(1) Complete whole garment

In the whole garment system one individual makes the entire garment from cutting the cloth to sewing and pressing the garment. The garment is ready for dispatch once the operator completes the final operation. This type of system is used in few places, which are engaged in custom-whole sale. They are normally high priced and exclusively made for a particular customer. They are limited in number and distribution; normally about 10 – 20 garments are made.

(2) Departmental whole garment.

The departmental whole garment system is also used by custom wholesale manufacturers as well as high price or better dress manufacturers. In the departmental whole garment system one individual does all the work with the equipment allocated to a department. For example on person does all the cutting work in cutting department, second person does all the sewing work in sewing department, third person does the pressing and packing work. The workers in this system may use more than one equipment to complete his/her job.

INDIVIDUAL SYSTEM

Advantages

1. This system is more effective when a very large variety of garments have to be produced in extremely small quantities.

2. In Individual piece rate system the operators will do with full involvement. To finish more pieces, to earn more money.

3. Operator will be specialized in his own working area.

4. As the pay depends upon the complication of the operation, the operator will try to finishes the complicated operation also without any difficulties.

5. The Work in Progress (WIP) is reduced, at a time one cut garment to one operator and so the amount as inventory is reduced.

Disadvantages

1.   Highly skilled labors are used, so the cost of labors is high.

2.   The operator is more concerned on the number of pieces finished rather than the quality of work.

3.   Productivity is less due to lack of specialization.

4.   For long run / Bulk Quantity of same style is not effective in this system.

SECTION OR PROCESS SYSTEM – GROUP SYSTEM

This is a development of the making through system, with the difference that the operators specialize in one major component and sew it from beginning to end. For example, an operator specializing in fronts would assemble the front, set the pockets, etc. and perform all the operations required to finished that particular component.

The sewing room would have a number of sections, each containing versatile operators capable of performing all the operations required for a specific component. The sections are built according to the average garment produced, and include:

- Pre-assembling (the preparation of small parts)
- Front making
- Back making
- Main assembly (closing, setting collars and sleeves etc.)
- Lining making

- Setting linings

- Finishing operations (buttonholes, blind-stiching etc.)

Some features of this system are:

- Supervision: This is more involved in production and has to ensure the correct movement of work from section to section.

- Labour: As labour with varying levels of skill can be used, the system is somewhat cheaper than making through.

- Quality: Due to the various levels of operator levels of operator skill, in-process quality contol has to be very thorough.

- Productivity: Somewhat higher than the making through system, but still requires operators to perform simple operations within the context of their specialization.

- Throughput Time: This is longer due to the larger quantities of work in process

- Layout: This would be planned according to the process flow of the average garment produced by the factory. All 'on-time' special operations would be performed off-section on machinery not a permanent part of a section.

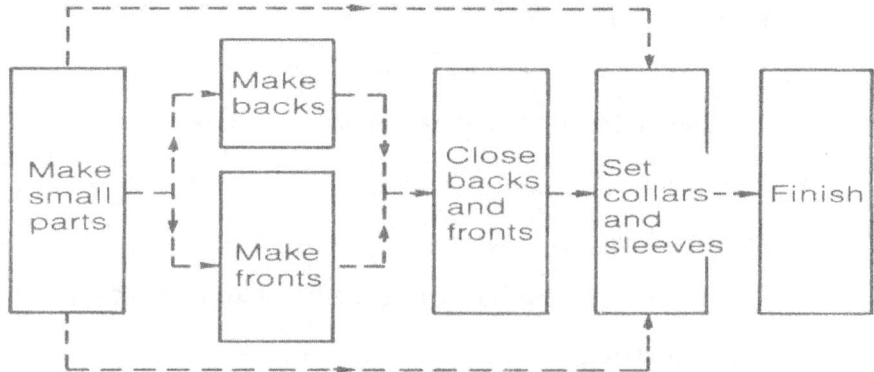

All in all, this is a very efficient system for producing a variety of styles in reasonable quantities. Above figure shows a typical layout and work flow for this type of system.

GROUP SYSTEM

Advantages

1. As the labour of all levels i.e. semi skilled, skilled, trainee can be used in this system and the labor cost is less compared with individual system.

2. Productivity is higher compare to individual system, because of use of special machine and all types of labour.

3. This system is very efficient for producing a variety of styles in reasonable quantities.

4. Automation and Specialization can be done.

5. Absenteeism and Machine breakdown problems will not cause serious problems.

Disadvantages

1. All the level of operators involved in the work, so the quality of garment should be strictly maintained.

2. Even though productivity is high still the highly skilled operators required to perform simple operation within the section.

3. Group of people involved in each section and so we require more WIP which increases the inventory cost.

4.   As this is not a bundling system, there is more chances to mix up of lost, shade variation, sizes, so quality and production will be affected.

PROGRESSIVE BUNDLE SYSTEM – BATCH SYSTEM

This system is exactly what its name implies, a system whereby the garments are gradually assembled as they move through successive sub-assembly and main assembly operations in bundle form. The principles of this system are:

- The various sections are positioned according to main operation sequence, with each section having a layout according to the sequence of operations required to produce a particular component. For example, the sleeve section could contain the following sequence of operations:

      1. sew seam

      2. press seam

      3. blind-stitch hem etc.

The amount of machinery for each operation would be determined by the output required.

- A work store is positioned at the start and end of every section of these buffers are used to store work received from a preceding operation, and to hold work completed by that section.

- Due to these work stores or buffers, each section is not directly dependent on the preceding section, but can absorb slight variations in output via the stocks held within the section.

The major feature if this system are:

- **Supervision**: This is not just involved in ensuring the correct movement of work between sections, but is concerned with the movement and balance of work within the section.
- **Labour**: This can be various skill and cost levels because the more complex operations are broken down into a series of small and relatively simple operations.
- **Quality**: In-process quality control is now involved in inspecting individual operations rather than completed components.
- **Productivity**: Due to the breakdown of operations and the possibility of introducing speciality machines, productivity is relative higher than that of more other systems.
- **Throughput Time:** This is considerably increased due to the amount of work stored at the beginning and end of each section.
- **Layout**: This is particularly important because the end of one section must be positioned adjacent to the start of the section, which performs the following series of operations.

The progressive bundle system, while being somewhat cumbersome in operation and requiring large quantities of work in progress, is probably one of the most stable systems as regards output. Unless there is serious absenteeism or prolonged special machine breakdowns, most of the usual hold-ups can be absorbed because of the amounts of work in progress. Balancing and the change over to new styles is also somewhat simplified, due to the amounts of work held in reverse. When properly managed the progressive bundle system is versatile and efficient. Figure 15.2 shows a typical layout and work-flow for this system.

BATCH SYSTEM

Advantages

1. Labours of all levels i.e. unskilled, skilled, semi-skilled labours are involved in this system where the operations are broken into small simple operation. Hence the cost of labour is very cheap.
2. Here the quantity of each component are checked during the individual operation itself, so the quality is good.
3. The components are moved in bundles from one operation to next operation, so there is less chance for confusion like, lot mix-up, shade variation, size variation etc.

4. Specialization and rhythm of operation increases productivity.

5. As the WIP is high in this system – This is stable system. Because of the buffer, the breakdown, absenteeism, balancing of line, change of style can be easily managed.

6. An effective production control system and Quality control system can be implemented.

   a. Time study, method study techniques.

   b. Operator training programme.

   c. Use of material handling equipments, such as centre table, chute, conveyor, trolley, bins etc.

7. Bundle tracking is possible, so identifying and solving the problems becomes easy.

Disadvantages

1. Balancing the line is difficult and this problem is solved by effective supervisor.

2. Proper maintenance of equipment and machinery is needed.

3. Proper planning requires for each batch and for each style, which takes lot of time.

4. Improper planning causes labour turnover, poor quality, less production etc.

5. Increase in WIP in each section increases the inventory cost.

6. Planned and proper layout should be made to make the system effective i.e. smooth flow of material.

7. Variety of styles, less quantity is not effective in this system.

8. Shuttle operators and utility operators needed in every batch to balance the line effectively.

Straight-line or 'synchro' system

As its name suggests, this system is based on a synchronized flow of work through each stage of producing a garment. Time-synchronization is the most important factor of this system because the flow of work cannot be synchronized if there are considerable variations in the standard times allowed for all the operations performed on the line. For example, if one operation has a value of 1.5 minutesm then all the other operations in the line must have the same, or a very close, value. The manipulation required to balance the standard time for each operator can sometimes lead to illogical combinations of whole or part operations which are not always conducive to the overall efficiency of individual operators.

The synchro system by its very nature is rigid and particularly vulnerable to absenteeism and machine breakdowns. At all times reserve operators and machines must be available to fill the gaps. In addition, this system requires a sufficient volume of the same type of garment to keep the line in continuous operation.

The basic feature of this system are:

- **Supervision**: Due to its rigidity, supervisors are very much concerned with keeping the line in balance at all times. Every minor delay could have serious repercussions.

- **Labour**: The operators require relatively high skill levels due to the combination of different operations which sometimes have to be performed in order to maintain a time balance between the operations in the line.

- **Quality**: In-process quality control must be more alert and intensive because hol-ups caused through quality problems can stop the line in a matter of minutes.

- **Productivity**: All things being equal, productivity levels can be very high due to the regular pace of the successive operations.

- **Throughput Time**: This is very short as a result of the quantity of work in process. There are no intermediate work stores other than the bundles awaiting the next operation.

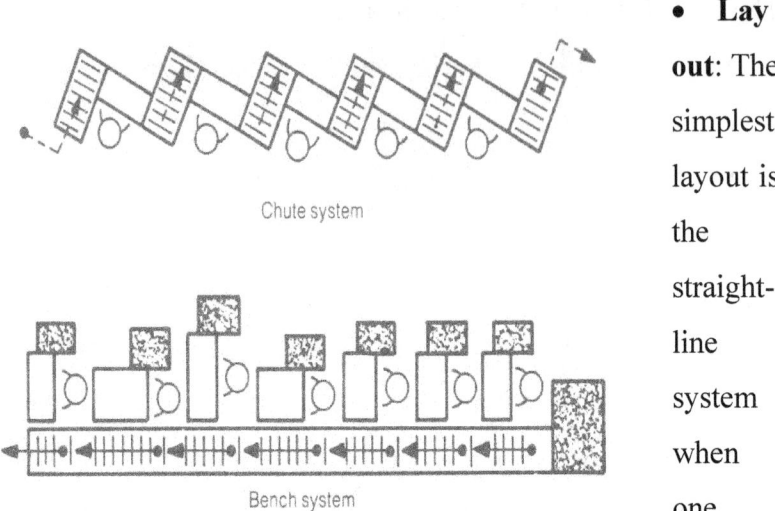

Chute system

Bench system

- **Lay out**: The simplest layout is the straight-line system when one operator is seated behind or opposite the next one. Work can be fed from one operator to the other by gravity chutes or by simply pushing bundle to work in the right direction along the bench. To be effective, this system requires:

- Volume production
- Accurate line balancing
- Skilled supervision
- Reserve operators
- Reserve machinery and equipment

Layout for Full Sleeve Shirt – Batch System

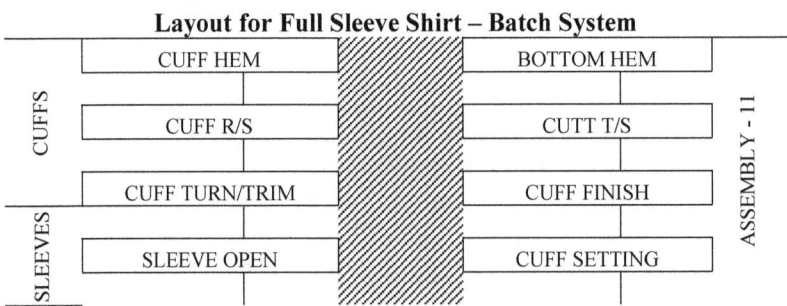

CUFFS	CUFF HEM		BOTTOM HEM	ASSEMBLY - 11
	CUFF R/S		CUTT T/S	
	CUFF TURN/TRIM		CUFF FINISH	
SLEEVES	SLEEVE OPEN		CUFF SETTING	

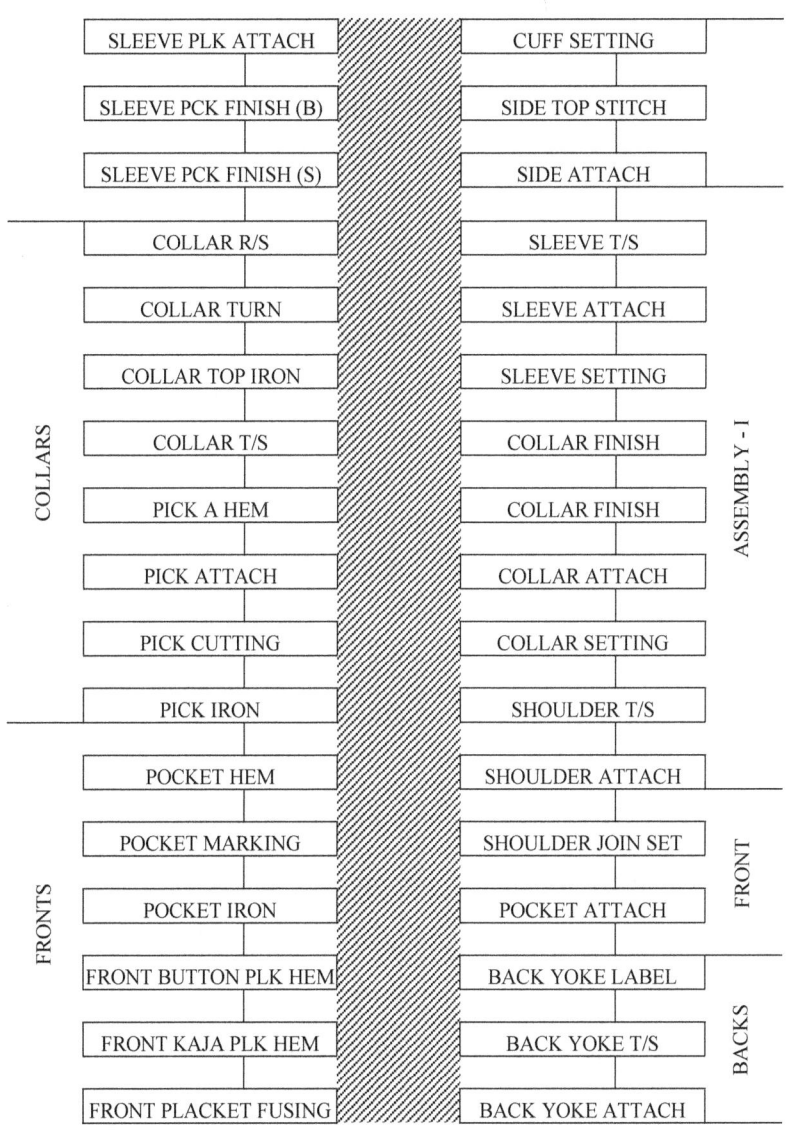

UNIT PRODUCTION SYSTEM (UPS)

As a mechanical system this has been in use for many years, but a major advance was made in 1983 when computers were first used to plan, control and direct the flow of work through the system.

The essential feature of this type of system are:

1. The unit of production is a single garment and not bundles.

2. The garment components are automatically transported from work station to work station according to a pre-determined sequence.

3. The work stations are so constructed that the components are presented as close as possible to the operator's left hand in order to reduce the amount of movement required to grasp and position and component to be sewn.

The operational principles are as follows:

All the components for one garment are loaded into a carrier at a work station specially designed for this purpose. The carrier itself is divided into sections, with each section having a quick-release clamp which prevents the components from falling out during movement through the system. When a batch of garments has been loaded into carriers they are fed past a mechanical or electronic device which records the number of the carrier and addresses it to its first destination. Some of the more intelligent systems address the carriers with all the destinations they will have to pass through to completion.

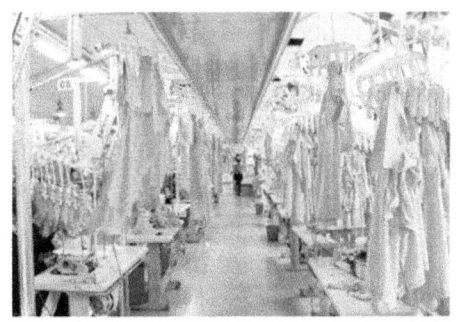

Fig. UPS Work Station

The loaded carriers are then fed on to the main powered line, which continually circulates between the rows of machines. This main, or head, line is connected to each work station by junctions which open automatically if the work on a carrier is addressed to that particular station. The carrier is directed to the left side of the operator and waits its turn along with the other carriers in the station. (Fig. UPS Work Station)

When the operator has completed work on one carrier, a push button at the side of the sewing machine is pressed and this actuates a mechanism, which transports the carrier back to the main line. As one carrier leaves the station, another is automatically fed in to take its place. When the carrier leaves the station it is recorded on the data collection system, and then addressed to its next destination.

- The marked advantages of this type of system are:
- Bundle handling on the part of the operator is completely eliminated
- The time involved in the pick-up of work and its disposal is reduced to the minimum possible.

- Output is automatically recorded, thus eliminating the necessity for the operator to register work.

- The computerized systems automatically balance the work between stations performing the same operation.

- Up to 40 styles can be produces simultaneously on one syste

Factors

- **Supervision:** Freed to work with the operators.
- **Labour**: All grades and all rates.
- **Quality control**: In-process inspection stations are built into the line and the inspector is able to return faulty work via the system to the operator concerned.
- **Throughput line:** This is measured in hours instead of days because of the low amounts of work in hand.
- **Productivity**: High because the operators are working in a 'paced' environment which enables them to develop faster working rhythms.
- **Layout**: Can be of any form suitable to the available area.

Unit Production System requires substantial investments, which are not always justified by conventional pay-back calculations. Apart from the measurable tangible benefits, UPS also have many intangible benefits such as a more orderly and controlled flow of work, and the ability via the control computer of simulating the production situation some time in advance. These intangibles are difficult to measure, but in themselves make a very positive contribution to the overall viability of the unit.

All things considered, unit production systems have major advantages over all the other manual and the mechanical systems used for the mass production of clothing. Most importantly, they provide a clothing factory with the capability to respond quickly to any changes, which might occur. In the fast moving fashion business, this is essential.

UNIT PRODUCTION SYSTEM

Advantages

1. Bundle handling completely eliminated.
2. The time involved in the pick up and disposal is reduced to minimum.
3. Output is automatically recorded, eliminates the operator to register the work.
4. The computerized systems automatically balance the work between stations.
5. Upto 40 styles can be produces simultaneously on one system.

Disadvantages

1. Unit production system requires high investments.

2.   The pay back period of the investment takes long time.

3.   Proper planning requires to be effective.

Quick response sewing system

This system was first developed in Japan to enable quick responses to be made to market changes, especially when orders for individual styles were in small lots. Each work station is equipped with two or four machines and the operator will take the garment through the required operations, including pressing, before it is transported to the next work station.

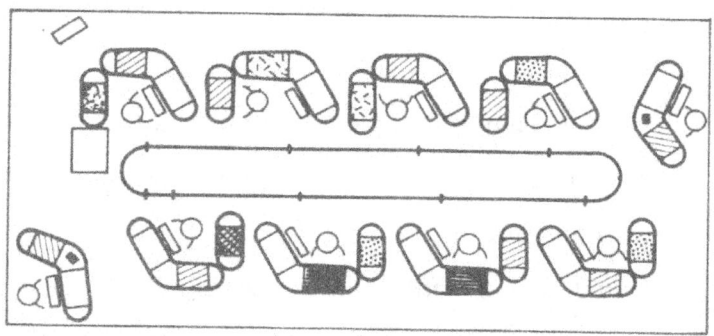

Fig. Quick response system layout

Some of the basic machinery is duplicated in different stations and if there is a bottle-neck in one section the over-load is automatically transported to other stations where operator capacity is available.

All the parts of one garment are loaded into a hanging clamp attached to the trolley and in theory, there should only be one garment at each work station. Work is transported by a computer controlled, overhead trolley system and each station has an individual controller which provides the operator with information on the style being worked on. This information comes from an information card which accompanies each trolley.

A less sophisticated version of QRS uses a wheeled trolley which contains the components for one garment and is pushed along the floor from operator to operator.

Another feature of QRS is that all the operators work in a standing position so that they can move quickly from one machine to another within their own work station. Machine heights are adjusted accordingly and touch-pads and knee-pads controls are used instead of conventional foot pedals.

Factors

- **Supervision**: Freed to work with the operators
- **Labour**: Of necessity the operators must be highly skilled in the operation of all the different machines in one work station.
- **Quality**: In-process inspection stations are built into the line and the inspector is able to return faulty work via the system to the operator concerned.
- **Productivity**: This is very high because the operator handles the garment once only for a number of operations, instead of once for each operation.
- **Throughput time:** As there are so few garments on the line throughput time is extremely short, which is the objective of this system.

- **Layout**: A typical unit would have eight work stations arranges around the transport system.

There is no doubt that this type of system is one of the best answers to the garment production revolution which is becoming more apparent every day. Fashion changes are becoming more frequent and as a consequence order lots are proportionately smaller. A production system which enables changeovers to be made in the minimum of time is ideally suited to this new and dynamic situation.

Individual Part System:

In this individual incentive systems (piecework) sewing machine operators, finishers and presses are paid on the piecework rate. That is, they are paid a set amount for each operation that they complete, rather than by the hour. Rates vary with the difficulty of operation. As proof of work completed, the operator signs the identification ticket or removes one segment of it. Actually, most companies pay a guaranteed wage, with piecework acting as an incentive for operators to work faster and therefore earn more.

Advantages of individual part system:

- The worker / the operators will do with full involvement to finish more pieces, the
- Operator will be specialized on his own working area
- As the pay is also made based on the complication of the operation, the operator will try to finishes the complicated operation also

Disadvantages of individual part production system:

- The operator is more concerned on the no. of pieces finished rather than the quality of work.

PRINCIPLES OF CHOOSING A PRODUCTION SYSTEM

The choice of best apparel production system will depend on the product and policies of the company and on the capacities of manpower. Where style changes are frequent and lot sizes are small, it may be advantageous to use skilled labour who can make whole garment and use one of the whole garment system. As the lot size increases it is advisable to use section production system. The sub-assembly system is superior to the progressive bundle system as it takes less time. That is the processing time for a garment in both system is same but sub-assembly system has less waiting or temporary storage time. However the space requirement, machinery requirement and labour costs are high for sub-assembly system.

In most cases the choice of a production system depends on the cost of the inventory-in process. Inventory-in process is the total number of garments in the production line. This consists of all garments being processed at sewing machines, under inspection and in temporary storage between operations. When material, labour, space and interest costs are high, synchronized sub-assembly system which yields the least possible in-process inventory is more suitable.

One of the aims of any production system is to make total production time as minimum as possible. This automatically reduces inventory cost to a minimum. Sub-assembly system provides many opportunities to economize on temporary storage and transportation space and time.

EVALUATION OF PRODCUTION SYSTEM

Any production system has four primary factors, which make up the system.

Processing Time + Transportation Time + Temporary Storage Time + Inspection Time = Total Production Time

Processing time is sum total of working time of all operations involved in manufacture of a garment. Transportation time involves the time taken to transport semi-finished or finished garments from one department to another or from one operation / machine to another. Temporary storage time is time during which the garment / bundle is idle as it waits for next operation or for completion of certain parts. Inspection time is time taken for inspecting semi-finished garments for any defects during manufacturing or inspecting fully finished garments before packing. The main aim of any production system is to achieve minimum possible total production time. This automatically reduces in-process inventory and its cost. The sub-assembly system reduces temporary storage time to zero by combining temporary storage time with transportation time. No definite answer can be given as to which is the best, as it depends on garment style, specifications, machinery and manpower and manufacturing policies.

Product Flow Chart

This is a graphical representation of operation, transportation, inspection, delays and storage occurring during production. This also gives the information regarding distances moved and time required for different items.

Process chart symbols

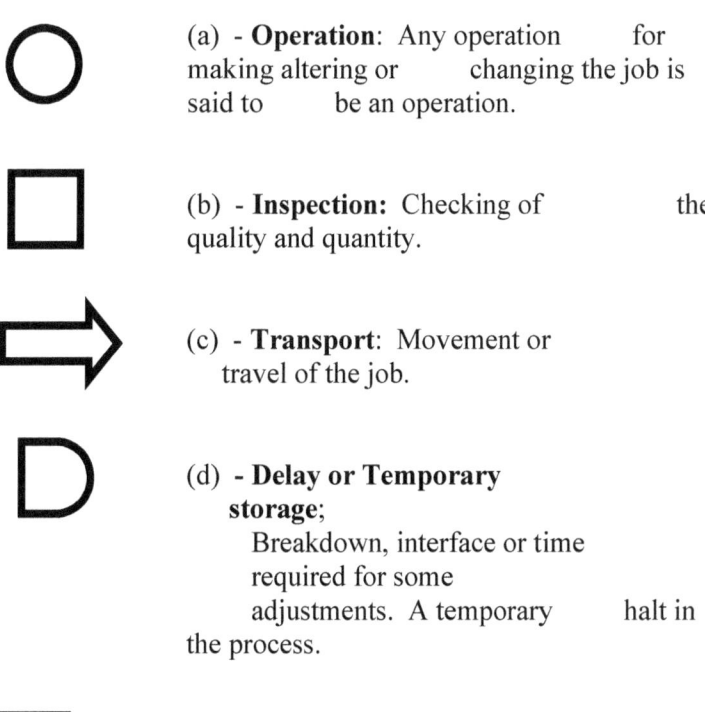

(a) - **Operation**: Any operation for making altering or changing the job is said to be an operation.

(b) - **Inspection:** Checking of the quality and quantity.

(c) - **Transport**: Movement or travel of the job.

(d) - **Delay or Temporary storage**;
Breakdown, interface or time required for some adjustments. A temporary halt in the process.

(e) - **Storage**: Keeping, holding and storing the job and other things.

T-shirt

T-SHIRT

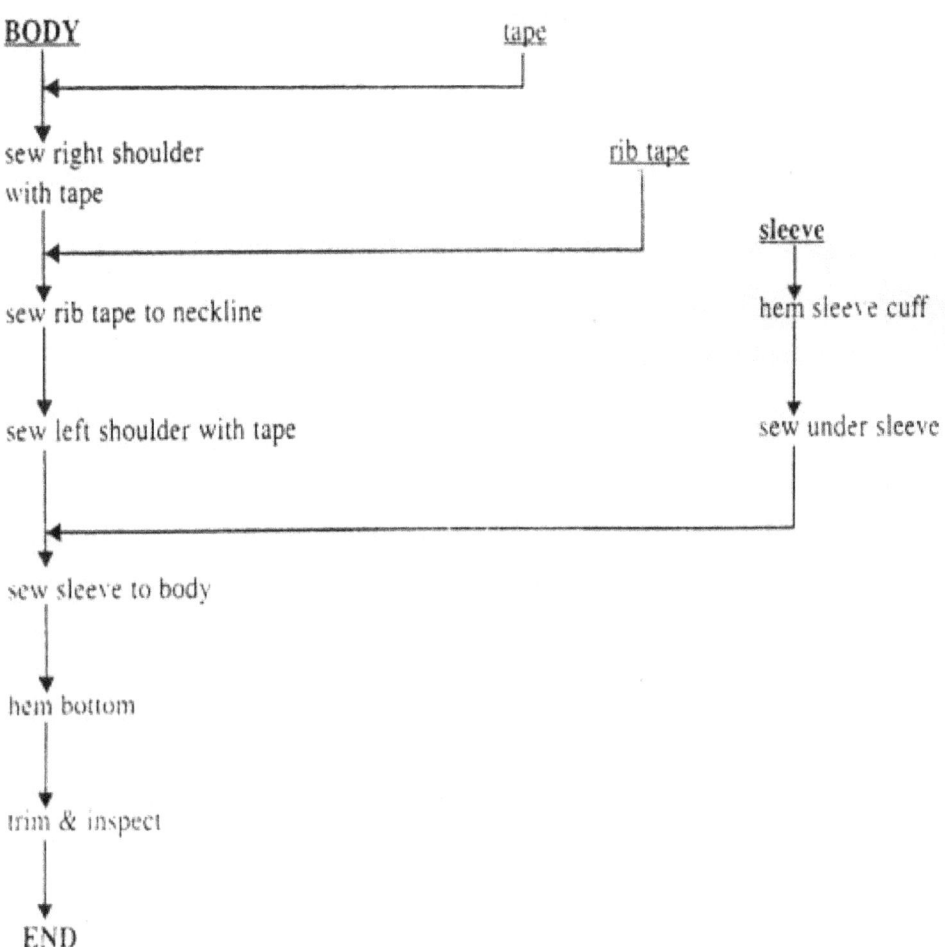

BODY   tape

sew right shoulder
with tape   rib tape

                                               sleeve

sew rib tape to neckline   hem sleeve cuff

sew left shoulder with tape   sew under sleeve

sew sleeve to body

hem bottom

trim & inspect

END

Balancing

Balancing is the technique of maintaining the same level of inventory at each and every operation at any point of time to meet the production target and to produce garments of acceptable quality.

- In operation breakdown we maximum try to equalize the std. time
- But still there will be the difference in the std. time which leads to work in progress.
- So, we try to set the flow through each operation to be similar as possible.
- Checking from time to time to see how things are going and then making adjustments to even out the flow again.

This process is called Balancing.

Theoretical Balance

Operation break down: Jobs must be broken down into operations of equal size.

Alternatives

S.No	Operative in parallel	Long operations performed by 2 or more people	Improved flexibility
1	Operative in series	Long operations split	Greater specialization
2	Method / construction	One or more elements combined together	May be increase in

	change		efficiency
3	Work place improvement	Work study & capital investment concentrated at bottle necks	Reduce manufacturing time

WIP (Work in Process)

Any garment which is unifinished and is in process of getting finished constitutes WIP.

Need for balancing

1. Keeping inventory cost low
2. To enable the operator to work at the optimal pace.
3. To enable the supervisor to attend other problems.
4. To enable better production planning.
5. Balancing production Line results in on time shipments, low cost and ensures reorders.

Goals for balancing

1. Meet prodn. Schedule
2. Avoid the waiting time
3. Minimize over time
4. Protect operator earnings.

Rules for Balancing

1. Have between 3-5 bundles of WIP at each operation
2. Solve the problem before they become any larger

3. Meet production goals by keeping every operator working at maximum capacity and make sure he has constant feeding to ensure his capacity is high.

How to balance the time

1. Know work available at the start of the day.
2. Plan transfer needed to compensate for any known absenteeism.
3. Check attendance at the start of the day
4. Make additional assignments to compensate unexpected absentees.
5. Make periodic checks during the day to check production.

Points to be noted when making balancing

1. Meet production target by usage of
   a. Regular operators
   b. Utility operators
   c. Shuttle operators
2. Work flow should be constant throughout all operations
3. Avoid over time
4. Determine human resource
5. Check absence daily
6. Assign utility shuttle operators based on need
7. Update daily production every two hours.

www.ingramcontent.com/pod-product-compliance
Lightning Source LLC
Chambersburg PA
CBHW061238180526
45170CB00003B/1351